Manual of

Axiomatic Set Theory

The Appleton–Century Mathematics Series

Raymond W. Brink and John M. H. Olmsted, Editors

Manual of

Axiomatic Set Theory

FRANK D. QUIGLEY

Tulane University

NEW YORK

APPLETON–CENTURY–CROFTS
EDUCATIONAL DIVISION
MEREDITH CORPORATION

Library of Congress Card Number: 72–110246

730-1

PRINTED IN THE UNITED STATES OF AMERICA
390–72360–6

PREFACE

The purpose of this book is to train students in the technique of abstract mathematical proof, while at the same time providing some basic information about sets. The text is primarily a list of definitions and theorems which presents some of the fundamental facts about the theory of sets, particularly cardinal and ordinal numbers. The subject matter is broken up into pieces sufficiently small that proofs can be provided by the students.

The development is based on one of the standard axiom systems for set theory, and once the axiom of choice is introduced, it is used freely, as is customary in modern mathematics. In general, the theorems become more difficult as the book proceeds, and some of the particularly difficult ones have been starred.

This book is based on notes written for use by first-year graduate students at Tulane University, and proved to be quite effective. The author has used the notes successfully with average seniors [mostly Chapter 1] and with very bright seniors [the whole book].

F.D.Q.

New Orleans, Louisiana

v

CONTENTS

AXIOMS FOR SET THEORY

Listed here are the axioms for our set theory. Explanations are given in the book as the various axioms are introduced. The terms ' class ' and ' set ' and the relation $\in$ are undefined. Equality is taken to mean identity of objects, and the validity of free substitution of equals for equals is assumed, though we do not explicitly formulate it as an axiom.

AXI $A = B$ *iff* (for all z, $z \in A$ iff $z \in B$).

AXII   Let P be any property. Then there exists a class A such that $x \in A$ *iff* Px and x is a set.

AX1 $\varnothing$ is a set.

AX2   If X and Y are sets, then so is $\{X, Y\}$.

AX3   If X is a set, then so is $\cup X$.

AX4   If X is a set, then there exists a set Y such that $A \in Y$ *iff* $A \subset X$.

AX5   If F is a generalized function, and if X is a set, then $F[X]$ is a set.

AX6 ω is a set.

AXC   If X is a set such that $\emptyset \notin X$, then there exists a function f with domain X such that $f(x) \in x$, for all $x \in X$.

GCH   [Generalized Continuum Hypothesis] $2^{\aleph_a} = \aleph_{a+1}$.

AXT   [Tarski's Axiom] For every set N there exists a set M with the following properties:

1^0 $N \in M$;

2^0 If $X \in M$ and $Y \subset X$, then $Y \in M$;

3^0 If $X \in M$, then $p(X) \in M$; and

4^0 If $X \subset M$ and if X is not equipotent with M, then $X \in M$.

[Axiom 6 and Axiom 4 imply Axiom 1; AXT implies AXC.]

Manual of

Axiomatic Set Theory

Chapter I

Elementary set theory

We shall make no attempt to introduce a formal language but shall be content with the following definition:

Definition 1.1

Certain logical conventions are required:

The mathematical 'if p, then q' is true except when p is true but q is false.

'p or q' is inclusive, that is, true except when p and q are both false.

The symbols $\exists$ and $\forall$ mean respectively 'there exists [is]' or 'there exist [are]', and 'for all'.

'p if and only if q' will be abbreviated 'p iff q'. It is true just when p and q are both true or both false. The symbol | [or ∋] means 'such that'.

The theory of sets deals with objects, which we shall call *classes*, for which a certain binary relation $\in$ called *membership* is defined. Thus, if A and B are classes, then '$A \in B$' [read A is a member of, belongs to, is in B] is meaningful. [We use single quotes to name the object within the quotes. Thus, A is a class,

1

but 'A' is the first letter of the Roman alphabet, and '$A \in B$' is the statement that $A \in B$.]

From this point on in this book *the only objects under consideration will be classes*, and such phrases as 'for all A' will mean 'for all classes A'. In short, everything is a class.

A basic property of classes is that they are entirely determined by their members: if two classes have precisely the same members, then they are identically the same object.

AXIOM I $A = B$ *iff* (for all z, $z \in A$ iff $z \in B$).

We interpret ' $=$ ' as the sign of *identity* of two objects. Thus '$A = B$' means that A and B are the same object. This implies that in all contexts either of these two names, 'A' and 'B', for the object may be substituted for the other. In particular, if such a substitution is made in a theorem, the result is again a theorem. [Within a specific formal language this principle of substitution can be made precise, using, in fact, Axiom I as the definition of equality. This, however, is a long and delicate procedure which we prefer to avoid, so that Axiom I is an axiomatic property of $\in$, and substitution of equals for equals remains an intuitive procedure.]

Since a class A is fully determined by its members, it will also be determined by any necessary and sufficient condition for membership in it, that is, A will be determined by any property P such that $x \in A$ iff x has property P. [We shall write ' Px ' for ' x has property P '.] Unfortunately, to admit the existence of classes corresponding to arbitrary properties leads to paradoxical complications. Consider the well-known example of Russell's: Let A be the class of all classes B such that $B \notin B$ [read, B is not a member of B], that is, $B \in A$ iff $B \notin B$; taking $B = A$, we get $A \in A$ iff $A \notin A$, which is contradictory.

There is a variety of ways to get around this difficulty, and we shall discuss one which is commonly used and which is, from a practical mathematical point of view, perhaps somewhat less artificial, or at least more convenient, than most. In order to do mathematics one must suppose that the common mathematical operations on classes—subsequently introduced—do not lead to trouble when applied to classes which arise naturally in mathematics, though trouble may come if they are applied to such objects as Russell's class. This suggests that one might break classes up into two kinds: the OK classes on which operations may be performed freely, and the others, which may be convenient to have around, but which must be watched. The OK classes we shall call *sets*; the others are often called *proper classes*.[1] The operations to be introduced will be freely applicable to sets and will yield sets, but will not, in general, be applicable to proper classes at all.

Intuitively, sets are those classes that are capable of membership. We make the following definition:

Definition 1.2

A is a set *iff* there exists B such that $A \in B$.

We now give a general axiom for the existence of classes.

AXIOM II   Let P be any property. Then there exists a class A such that $x \in A$ *iff* Px and x is a set.

Definition 1.3

We write '$\{x|Px\}$' for the class of Axiom II. Thus $y \in \{x|Px\}$ *iff* Py and y is a set.

If 'Px' is Russell's property '$x \notin x$' and $A = \{x|x \notin x\}$, then $A \in A$ *iff* $A \notin A$ and A is a set, which is no longer contradictory but merely implies that A is not a set.

---

[1] This use of the term 'proper' should not be confused with its meaning in Definition 1.4. Some mathematical terms have several meanings.

The foregoing distinction between class and set is a technical one which is only observed in mathematical writing when necessary to avoid possible contradiction; for example, one would speak of the class of all topological spaces [it is a proper class], not the set of all such.   Similarly, the category of abelian groups and homomorphisms is not a set, but it is a very useful class. Normally, however, such terms as set, class, collection, aggregate, family, etc. are used interchangeably in mathematical writing. The mathematician only becomes cautious when he hears contradiction " whiffling through the tulgey wood."

Definition 1.4

$A \subset B$ [$B \supset A$] *iff* for all z, $z \in B$ whenever $z \in A$.   [Read, A is included in B, A is a subclass of B, B includes A.]

A is a *proper* subclass of B *iff* $A \subset B$ but $A \neq B$.

$p(B) = \{A | A \subset B\}$.   [The class of all sub*sets* of B, cf. Definition 1.3.]

$\emptyset$ is the class with *no* members.   [The *empty* or null class.]

Definition 1.5

1^0 $A \cup B = \{z | z \in A \text{ or } z \in B\}$.   [The *union* of A and B.]

2^0 Let X be a class of sets.

$\cup X = \{z \mid \text{for some } A \in X, z \in A\}$.

$\cap X = \{z \mid z \in A, \text{ for all } A \in X\}$.

[If $X = \emptyset$, then $\cup X = \emptyset$.   However, $\cap X$ in this case becomes the class of all *sets*, cf. Definition 1.3. Frequently, in a particular context when X is understood to be a set of subsets of a given set Z, one adopts the convention that $\cap X = Z$ whenever $X = \emptyset$.]

3^0 $A - B = \{z \mid z \in A \text{ and } z \notin B\}$. [The *relative comple-ment* of B in A. It is also annotated $A \setminus B$ or $A \sim B$.]

Theorem 1

Let A and B be subclasses of X. Then the following state-ments are equivalent:

1^0 $A \subset B$,

2^0 $A \cap B = A$,

3^0 $A \cup B = B$,

4^0 $(X - B) \subset (X - A)$,

5^0 $A \cap (X - B) = \varnothing$,

6^0 $(X - A) \cup B = X$.

Definition 1 . 6

$\{a_1, \ldots, a_n\}$ is the class whose elements are the sets $a_1, \ldots, a_n$.

(a, b) is a class characterized by the property that

$$(a, b) = (c, d) \text{ iff } a = c \text{ and } b = d.$$

[It will be proved in Theorem 4 that the class (a, b), which is called the *ordered pair* of the sets a and b, may be *defined* as follows: $(a, b) = \{\{a\}, \{a, b\}\}$.]

$A \times B = \{(a, b) \mid a \in A \text{ and } b \in B\}$.

[The *Cartesian product* of A and B.]

[We have defined $\{a\}$ only if a is a set. It is perfectly possible to define $\{a\}$, when a is a proper class, in various ways so that it comes out Ø or the class of all sets or even *a* itself. But these are all in effect conventional meanings of no mathematical use, though they would be convenient in a more formal development, simply to have {*a*} always defined. However, for the purposes of this book we shall leave {*a*}, and incidentally (*a*, *b*), undefined unless *a* and *b* are sets.]

Theorem 2

Let A, B, C, and D be classes. Then

1^0 $D - (D - A) = D \cap A$.

2^0 $A \cup B = B \cup A$, and $A \cap B = B \cap A$.

3^0 $A \cup (B \cup C) = (A \cup B) \cup C$, and

$A \cap (B \cap C) = (A \cap B) \cap C$.

4^0 $A \cap (B \cup C) = (A \cap B) \cup (A \cap C)$, and

$A \cup (B \cap C) = (A \cup B) \cap (A \cup C)$.

5^0 $D - (A \cup B) = (D - A) \cap (D - B)$, and

$D - (A \cap B) = (D - A) \cup (D - B)$.

6^0 $(A \times B) \cap (C \times D) = (A \cap C) \times (B \cap D)$.

Up to this point we do not know that any sets at all exist. A typical mathematical theory will assume axiomatically or implicitly that it is dealing with a sufficiently rich universe [= class, usually proper] of sets appropriate to that theory. For set theory itself we must assume outright that *something* is a set. The following axioms form a suitable beginning:

AXIOM 1   Ø is a set.

AXIOM 2   If X and Y are sets, then so is {X, Y}.

AXIOM 3   If X is a set, then so is $\cup X$.

Axiom 4   If X is a set, then there exists a set Y such that

$$A \in Y \; \textit{iff} \; A \subset X.$$

Theorem 3

Let X and Y be sets. Then so are the following:

$\{X\}$, any subclass of X, $p(X)$, and if $X \neq \varnothing$, $\cap X$; also

$\{\{X\}, \{X, Y\}\}$, $X \cup Y$, $X \cap Y$, and $X \times Y$.[2]

Theorem 4

Let a and b be sets and *define* $(a, b) = \{\{a\}, \{a, b\}\}$.
Then (a, b) is a set. Furthermore,

$\cup(a, b) = \{a, b\}$, $\cap(a, b) = \{a\}$,

$\cup\cap(a, b) = a$,

$\cap\cap(a, b) = a$,

$\cup\cup(a, b) = a \cup b$,

$\cap\cup(a, b) = a \cap b$,   and

$[\cap\cup(a, b)] \cup \big[\big(\cup\cup(a, b)\big) - \big(\cup\cap(a, b)\big)\big] = b.$

Corollary. $(a, b) = (c, d)$ if and only if $a = c$ and $b = d$.

Note that in the following definitions to avoid repetition of the word 'generalized' we shall write 'Relation' [with a capital 'R'] for 'generalized relation' and 'Function' for 'generalized function'. Those Relations and Functions which are *sets* will be annotated with the small 'r' and 'f', and without the 'generalized'.

---

[2] The last one should be proved after Theorem 4. It is included here to make the list complete.

RELATIONS AND FUNCTIONS

Definition 1.7

Let X and Y be classes, and let $X \times Y$ be their Cartesian product.

1^0 A *Relation R* on $X \times Y$ is any subclass of $X \times Y$. [That is, a Relation is a class of ordered pairs.]

x bears the Relation R to y *iff* $(x, y) \in R$;
other notation: xRy, and Rxy.

2^0 Let R be a Relation on $X \times Y$, and let $A \subset X$.

$\mathrm{pr}_1(R) = \{x \mid (x, y) \in R$ for some $y \in Y\}$.

$\mathrm{pr}_2(R) = \{y \mid (x, y) \in R$ for some $x \in X\}$.

$[\mathrm{pr}_1(R)$ is called the 1st projection [or *domain*] of R, and $\mathrm{pr}_2(R)$ is called the 2nd projection [or *range*] of R.]

$R[A] = \{y \mid (a, y) \in R$ for some $a \in A\}$.

$R^{-1} = \{(y, x) \mid (x, y) \in R\}$.   [The *inverse* of R.]

3^0 Let R and S be Relations.   Then

$R \circ S = \{(x, z) \mid$ there exists y for which $(x, y) \in S$ and $(y, z) \in R\}$.

$R \circ S$ is the composite of R and S.   [The twist in the definition is needed to make the composition of functions come out the ordinary way round.]

4^0 $R \mid A = R \cap (A \times pr_2(R))$.   [The *restriction* of R to A.]

Theorem 5

Let R, S, and T be Relations.   Then

1^0 $(R \circ S) \circ T = R \circ (S \circ T)$.

2^0 $R \circ (S \cup T) = (R \circ S) \cup (R \circ T)$.

3^0 $R \circ (S \cap T) \subset (R \circ S) \cap (R \circ T)$.

4^0 $(R^{-1})^{-1} = R$.

5^0 $(R \circ S)^{-1} = S^{-1} \circ R^{-1}$.

Definition 1.8

1^0 A *Function* F is a Relation such that for all x, y, z, if $(x, y) \in F$ and $(x, z) \in F$, then $y = z$.

2^0 If F is a Function, and if $x \in pr_1(F)$, then the *unique* y such that $(x, y) \in F$ is the *value* of F at x, and it is written $F(x)$.

3^0 Let F be a Function, and let B be a class.
F is *one-to-one* iff $F(x) = F(y)$ implies $x = y$ for all x, $y \in pr_1(F)$.
[Also called *injective*; F is an *injection*.]
F is *onto* B iff $pr_2(F) = B$.   [Also called *surjective*; F is a *surjection*.]
F is *into* B iff $pr_2(F) \subset B$.   [An F both one-to-one and onto is called *bijective* or a *bijection*.]

Theorem 6

Let F and G be Functions and R any Relation.

1^0 $F \circ G$ is a Function.

2^0 F^{-1} is a Function if and only if F is one-to-one.

3^0 $R[A \cup B] = R[A] \cup R[B]$, and
$R^{-1}[A \cup B] = R^{-1}[A] \cup R^{-1}[B]$.

4^0 $R^{-1}[A \cap B] \subset R^{-1}[A] \cap R^{-1}[B]$, but

$F^{-1}[A \cap B] = F^{-1}[A] \cap F^{-1}[B]$.

5^0 $R[A \cap B] \subset R[A] \cap R[B]$. If $A \subset B$, then $R[A] \subset R[B]$.

6^0 $F^{-1}[A - B] = F^{-1}[A] - F^{-1}[B]$.

7^0 If $A \subset B \subset pr_1(F)$, then $F \mid A = (F \mid B) \mid A$.

We insert at this point an axiom about Functions, even though its use can be avoided until the definition of the ordinal numbers.

AXIOM 5   If F is a generalized function, and if X is a set, then $F[X]$ is a set.

We now give another definition of union and intersection for classes of sets, which has a slightly different emphasis.

Definition 1 . 9

1^0 A class X is *indexed* by a class A and a Function f iff $pr_1 f = A$ and $pr_2 f = X$. We write $f(a) = X_a$, and $X = \{X_a, \ a \in A\}$.

Also $\cup X = \bigcup\limits_{a \in A} X_a$, and $\cap X = \bigcap\limits_{a \in A} X_a$, by definition.

Frequently the particular indexing Function f is not explicitly mentioned, and we might say: let X be indexed by a class A, or let $X = \{X_a, a \in A\}$.

2^0 If $B \subset A$, and if $X = \{X_a, \ a \in A\}$, then we write $\{X_b, \ b \in B\}$ for the class $pr_2(f \mid B)$ indexed by B and $f \mid B$. Observe that each a and each X_a is a set, since $a \in A$ and $X_a \in X$.

Theorem 7

For each $a \in A$, let $X_a \subset Y$, let $X = \{X_a,\ a \in A\}$, and let $B \subset A$. Let g be any Function, then:

1^0 $\cup \{X_b,\ b \in B\} \subset \cup X$.

2^0 $\cap \{X_b,\ b \in B\} \supset \cap X$.

3^0 $Y - \cup X = \cap \{(Y - X_a),\ a \in A\}$.

4^0 $Y - \cap X = \cup \{(Y - X_a),\ a \in A\}$.

5^0 $g^{-1}[\cup X] = \cup \{g^{-1}[X_a],\ a \in A\}$.

6^0 $g^{-1}[\cap X] = \cap \{g^{-1}[X_a],\ a \in A\}$.

Next we study Relations which have special properties.

Definition 1 . 10

R is a *Relation on X iff* $R \subset X \times X$. Write xRy to mean $(x, y) \in R$.

1^0 R is *reflexive iff* xRx, for all $x \in X$.

2^0 R is *symmetric iff* xRy implies yRx, for all $x, y \in X$.

3^0 R is *transitive iff* xRy and yRz imply that xRz, for all $x, y, z \in X$.

Theorem 8

Let R be a Relation on X.

1^0 Let $D = \{(x, x),\ x \in X\}$. Then R is reflexive iff $R \supset D$.

2^0 R is symmetric iff $R = R^{-1}$.

3^0 R is transitive iff $R \circ R \subset R$.

4^0 Call R *antisymmetric iff* xRy always implies (not yRx).

Then R is antisymmetric iff $R \cap R^{-1} = \varnothing$.

EQUIVALENCE RELATIONS

Definition 1 . 11

1^0 A Relation R on X is an *equivalence Relation iff*

(a) R is reflexive, symmetric, and transitive; and

(b) $pr_1 R = X$.

2^0 x is *equivalent* to y *iff* xRy.   We also write $x \equiv y$, $x \equiv_R y$, or $x \equiv y(R)$.

Example: Let X be the " set " of real numbers, and let R be the relation of ordinary equality.

We shall now determine the structure of all possible equivalence relations on a set X.   We begin with a definition and lemmas preliminary to the main result, which is Theorem 9.

Definition 1 . 12

1^0 A class X of sets is called *disjoint iff* for all $A, B \in X$, either $A = B$ or $A \cap B = \emptyset$.

2^0 If R is an equivalence Relation on X and $x \in X$, then $R[\{x\}]$ is called the *equivalence class* of x in X relative to R; that is, $R[\{x\}]$ is the class of all $y \in X$ such that xRy.   [NB: Frequently, when no confusion is likely, $R[x]$ is written, inaccurately, for $R[\{x\}]$, for typographic reasons.]

Example: The following familiar but nontrivial example should be borne in mind throughout the following.   To understand abstract mathematics it is always essential to have in mind some concrete examples.   Let X be the " set " of all integers; that is, all positive or negative whole numbers, and zero.   Let R be the set of all ordered pairs (x, y) of integers such that $x - y$ is a multiple—positive or negative—of some fixed integer m. Then 'xRy' is the ordinary congruence relation $x \equiv y \pmod{m}$ of number theory, and $R[\{x\}]$ is the congruence class of $x \pmod{m}$. [This example, like the previous one involving real numbers and others to follow, is quite informal, since integers, real numbers, etc. have not been defined formally.]

Lemma 9.1

Let R be an equivalence Relation on a class X. Then

1^0 $x \in R[\{x\}]$,   and

2^0 $R[\{x\}] \times R[\{x\}] \subset R$.

Lemma 9.2

Under the hypotheses of Lemma 9.1,

$$z \in R[\{x\}] \quad \text{iff} \quad R[\{x\}] = R[\{z\}].$$

The preceding lemmas have been formulated for generalized relations. The basic theorem, which follows, can not be so formulated, the difficulty being that without knowing that each $R[\{x\}]$ is a set, one can not assemble them into the class Y of this theorem.

Theorem 9

Let X be a class, and let R be a subclass of $X \times X$ such that $R[\{x\}]$ is a set for every $x \in X$. Then R is an equivalence Relation on X *iff* there exists a disjoint class Y of subsets of X such that $\cup Y = X$ and $R = \cup\{A \times A \mid A \in Y\}$. If X is itself a set, then any R on X and each $R[\{x\}]$ are sets.

Note that from this point on it will be assumed in theorems and definitions that X, Y, A, B, etc. are sets, unless otherwise stated.

When *new objects* are introduced and *called sets*, it may be appropriate to prove, using the axioms, that they are such; for example, X/R in Definition 1.13. Actually the next few theorems and definitions could be formulated for classes by using a hypothesis like that of Theorem 9, but there is no need to run the matter any farther into the ground.

Definition 1.13

1^0 Let R be an equivalence relation. The set $\{R[\{x\}],$ $x \in X\}$ is called the *quotient* [set] of X by [or relative to] R and is denoted by ' X/R '. Observe that equivalence on X becomes set equality on X/R in the sense that $x \equiv y(R)$ iff $R[\{x\}] = R[\{y\}]$.

2^0 The function F with domain X $[= pr_1 F]$ defined by the equation $F(x) = R[\{x\}]$ is called the *quotient mapping* [mapping = function], *canonical mapping*, or *natural mapping* of X onto X/R.

3^0 Let F be the canonical mapping of X onto X/R. A subset B of X such that $F \mid B$ is a 1–1 mapping of B onto X/R is called a *system of representatives* for the equivalence classes of R on X. [We postpone discussion of the existence of such a system of representatives for an arbitrary equivalence relation [small ' r '] until later.]

Theorem 10

Let G be a function with domain X and such that $pr_2 G \subset Y$.[3] Then the following are true:

1^0 If $S = \{(x, z), x, z \in X,$ such that $G(x) = G(z)\}$, then S is an equivalence relation on X.

---

[3] We use the notation ' $G : X \to Y$ ' in this situation.

2^0 Every equivalence relation R on X has the form just described in 1^0, for a suitable choice of G and Y.

[The relation S of Theorem 10.1^0 is called the equivalence *induced* on X by G.]

Definition 1.14

Let R be an equivalence relation on X.

1^0 A subset A of X is *saturated* by R *iff* for all $x, y \in X$, $x \in A$ and xRy imply $y \in A$.

2^0 A function g with domain X is *compatible* with R *iff* for all $x, y \in X$, xRy implies $g(x) = g(y)$.

Theorem 11

Let R be an equivalence relation on X, let $A \subset X$, and let F be the canonical mapping of X onto X/R. Then:

1^0 A is saturated by R iff $x \in A$ implies $R[\{x\}] \subset A$.

2^0 A is saturated by R iff $A = F^{-1}[F[A]]$.

3^0 If $B \subset X/R$, then $F^{-1}[B]$ is saturated by R.

4^0 If C is any subset of X, then $F^{-1}[F[C]]$ is saturated by R

5^0 Let Y be a family of subsets of X saturated by R; then $\cap Y$ and $\cup Y$ are saturated by R.

6^0 If A is saturated by R, then so is $X - A$.

Theorem 12

Let R be an equivalence relation on X, and let F be the canonical mapping of X onto X/R. If g is a function with domain X such that $pr_2 g \subset W$, then:

1^0 g is compatible with R iff for each $w \in pr_2 g$, $g^{-1}[\{w\}]$ is saturated by R.

2^0 g is compatible with R iff $F(x) = F(y)$ implies $g(x) = g(y)$, for all $x, y \in X$.

3^0 g is compatible with R iff there exists a function h from X/R into W such that $g = h \circ F$. The function h is unique.

Theorem 13

1^0 If $W \subset X$, and R is an equivalence relation on X, then $R \cap (W \times W)$ is an equivalence relation on W.

2^0 Let R and S be equivalence relations on X. Then $R \circ S$ is an equivalence relation on X iff $R \circ S = S \circ R$. In this case, $R \circ S \ [= S \circ R]$ equals $\cap\{T, \text{ such that } T \text{ is an equivalence relation on } X \text{ and } T \supset (R \cup S)\}$.

Definition 1 . 15

$R \cap (W \times W)$ is called the *restriction* of R to $W\,[W \subset X]$, or the relation *induced* on W by R. In this context, we understand W/R to mean $W/R \cap (W \times W)$, and call W/R the quotient [set] induced on W by R.

Chapter II

The axiom of choice

Before resuming the study of special relations, we shall consider the question raised in Definition $1.13.3^0$. Intuitively it seems likely that such a system of representatives exists. In fact, X/R is a disjoint family of subsets of X, and, for each $A \in X/R$, we need only choose one $a \in A$ [$A \neq \emptyset$, since R is an equivalence relation] and define B as the set of all such a. To be more precise, we might define a function f on X/R such that $f(A) \in A$, for each $A \in X/R$; then $pr_2 f$ would be a system of representatives for R. More generally, if X is any family of sets such that $\emptyset \notin X$, it seems reasonable that there should exist a function f with domain X such that $f(x) \in x$, for all $x \in X$. That the existence of such functions was implicitly a part of the reasoning employed in much late 19th- and early-20th-century mathematics was first observed by E. Zermelo in 1908. Mathematics in more recent years has made free and [usually] conscious use of the existence of such 'choice' functions, and we shall formulate their existence as an axiom.

AXIOM OF CHOICE [AXC]. If X is a set such that $\emptyset \notin X$, then there exists a function f with domain X such that $f(x) \in x$, for all $x \in X$.

COROLLARY.   In AXC both f and $f[X]$ are sets.

Modern mathematics has found that certain other principles, which imply and are implied by AXC, are often more convenient than that axiom itself.   At first glance these equivalent principles bear little resemblance to AXC, and several new definitions will be required even to state them.   First, however, we shall use AXC to extend Definition 1.7, which was made for two classes A and B, to an arbitrary indexed set of nonempty sets.

Definition 2.1

Let $X = \{X_a,\ a \in A\}$ be a set of nonempty sets, indexed by a set A.

1^0 The *cartesian product* of X, written $\Pi(X_a,\ a \in A\}$, or $\prod_{a \in A} X_a$, or even $\prod_A X$, equals by definition the *set* of *all* functions f with domain A such that $f(a) \in X_a$, for each $a \in A$.

[Note that repetitions among the X_a are allowed, that is, X_a may equal X_b even if $a \neq b$.]

2^0 Let $P = \Pi\{X_a,\ a \in A\}$.   For a *fixed* $a \in A$, the function pr_a with domain P and range X_a, defined by the equation $pr_a(f) = f(a)$, for each $f \in P$, is called the *a-projection* of P, or the projection of P on X_a; $pr_a(f)$ $[= f(a)]$ is the *a-coordinate* of f.

[In what follows the letters "OAXC" abbreviate "On the Axiom of Choice" and will be used in the statement of a theorem to show that its proof requires that axiom.]

Theorem 14 [*OAXC*]

Let $\{X_a,\ a \in A\}$ be a family of nonempty sets.   Then

$$\Pi\{X_a,\ a \in A\} \neq \varnothing;$$

in fact, each of the functions pr_a is *onto* X_a.

Theorem 15 [*OAXC*]

Any equivalence relation R on a set X has a system of representatives.

The next few definitions and theorems deal with various kinds of order relations.

Definition 2.2

1^0 A *quasi-ordered system* [QOS] is a pair (X, R), where X is a set and R is a reflexive and transitive relation on X.

2^0 A *partially ordered system* [POS] is a pair (X, R) where R is a reflexive and transitive relation on X [so that (X, R) is a QOS], and *furthermore* the relation 'xRy and $x \neq y$' is antisymmetric: that is, xRy and yRx imply $x = y$. [The latter condition is sometimes taken as the definition of antisymmetry.]

3^0 We say that X is quasi- ⟨partially⟩ ordered by R, or that R is a quasi- ⟨partial⟩ ordering of X. When the relation R is clear from the context, we may simply say that X is a quasi- ⟨partially⟩ ordered set.

Examples:

1^0 The cyclic ordering of the points on a circle C is a quasi-ordering but not a partial ordering; in fact any two points are comparable: $R = C \times C$.

2^0 Every equivalence relation is a quasi-ordering, but the relation '$x = y$' is the only equivalence relation which is a partial ordering.

3^0 The ordinary $\leqslant$ relation on the real numbers is a partial ordering.

4^0 Let X be the set of all real-valued functions of a real variable, and let

$$R = \{(f, g) \in X \times X \mid f(x) \leqslant g(x), \text{ all real } x\}.$$

Then (X, R) is a POS.

5^0 The relation $\subset$ on the set P of all subsets of any set X is a partial ordering of P.

6^0 Let X be the set of all [strictly] positive integers. Then the relation ' x divides y ' is a partial ordering of X.

[Note: Since it is suggestive, perhaps dangerously so, we frequently write $\leqslant$ for *any* quasi- $\langle$partial$\rangle$ order relation R on a set X; in fact, it is common to denote two different orders on two different sets by the same symbol ' $\leqslant$ ' if the context makes the meaning unambiguous. Thus $x \leqslant y$ means xRy; $x \geqslant y$ means yRx; $x < y$ means xRy but $x \neq y$; etc.]

Theorem 16

Let R be a relation on X, and let $D = \{(x, x) \in X \times X\}$. Then (X, R) is a POS iff

(a) $R \circ R = R$; and

(b) $R \cap R^{-1} = D$.

The next theorem shows that in a certain sense any quasi-ordering induces a partial ordering.

Theorem 17

Let (X, R) be a QOS, and let ' xSy ' be the relation ' xRy and yRx '.

1^0 S is an equivalence relation on X.

2^0 Let G be the canonical mapping of X onto X/S. *If* xRy, $x' \in G[x]$, and $y' \in G[y]$, then $x'Ry'$.

3^0 Define a relation $\bar{R}$ on X/S as follows:
for each $A, B \in X/S$, $A\bar{R}B$, *iff* there exist $x \in A$ and $y \in B$ such that xRy. Then $(X/S, \bar{R})$ is a POS.

Definition 2.3

Let $(X, \leqslant)$ be a QOS.

1^0 A subset Y of X is *totally ordered* [by $\leqslant$] *iff* for all $y, z \in Y$, either $y \leqslant z$ or $z \leqslant y$. If X itself has this property, we call $(X, \leqslant)$ totally ordered.

2^0 An element $x \in X$ is an *upper* $\langle lower \rangle$ *bound* for a subset Y of X *iff* $x \geqslant y$ $\langle x \leqslant y \rangle$ for all $y \in Y$.

3^0 An upper $\langle lower \rangle$ *bound* x for Y is a *least upper* $\langle greatest\ lower \rangle$ *bound* for Y *iff* every upper $\langle lower \rangle$ bound u for Y has the property $x \leqslant u$ $\langle x \geqslant u \rangle$. *In a POS,* a least upper $\langle greatest\ lower \rangle$ bound of Y is unique if it exists, and we denote it by lub Y or sup Y $\langle glb\ Y$ or inf $Y \rangle$.

4^0 An element $z \in X$ is *maximal* $\langle minimal \rangle$ *iff* $z \leqslant x$ implies $z \geqslant x$, for all $x \in X$ $\langle z \geqslant x$ implies $z \leqslant x$, for all $x \in X \rangle$.

Examples:

1^0 The real number system and its subsets are the examples that motivate the terminology.

2^0 Let P be the set of all subsets of X, partially ordered by $\subset$. If $Q \subset P$, then sup $Q = \cup Q$ and inf $Q = \cap Q$.

The following "fixed-point" theorem is fundamental in proving the equivalence of AXC to other principles.

*Theorem 18[1]

Let $(X, \leqslant)$ be a nonempty POS such that every totally ordered subset of X has a least upper bound.
If $f: X \to X$ is such that $f(x) \geqslant x$, for all $x \in X$, then there exists $w \in X$ such that $f(w) = w$.

Establish the theorem by completing the following outline of a proof, or otherwise.

Proof outline: Fix an element $b \in X$, to be used throughout the proof. Call a subset Y of X *admissible* iff it has the following properties:

1^0 $b \in Y$;

2^0 $f[Y] \subset Y$;   and

3^0 every least upper bound of a totally ordered subset of Y belongs to Y.

Let M be the set of all admissible subsets Y of X.

(a) $M \neq \varnothing$; $(M, \subset)$ is a POS, and $\cap M$ is the unique minimal element of M.

(b) If $x \in \cap M$, then $x \geqslant b$.
   Now let $P = \{x \in \cap M$ such that $(Y \in \cap M$ and $y < x)$ imply $f(y) \leqslant x\}$.

(c) $(x \in P$ and $x \in \cap M)$ imply either $z \leqslant x$ or $z \geqslant f(x)$. [Prove that for each $x \in P$, the set $A_x = \{z \in \cap M$ such that either $z \leqslant x$ or $z \geqslant f(x)\}$ is an admissible subset of $\cap M$; whence $A_x = \cap M$.]

[1] See Preface for an explanation of the asterisk.

(d) Using (c) when necessary, show that P is an admissible subset of $\cap M$; whence $P = \cap M$.

(e) Show that $\cap M$ is totally ordered, and then complete the proof of the theorem.

Theorem 19 [*OAXC*] [*Hausdorff Maximality Principle*]

Let the family T of all totally ordered subsets of a QOS $(X, \leqslant)$ be made into a POS $(T, \subset)$ under inclusion. Then T has a maximal element.

Proof outline: Using AXC and reasoning by contradiction, construct a function which contradicts Theorem 18.

The following result is one of the most important consequences of AXC [we shall shortly prove their equivalence], and it is in this form that modern mathematics makes its most deliberate use of AXC.

Theorem 20 [*OAXC*] [*Zorn's Lemma*]

If every totally ordered subset of a QOS $(X, \leqslant)$ has an upper bound, then $(X, \leqslant)$ has a maximal element.

Hint: Use Theorem 19 to prove this theorem.

[Note: A set satisfying the hypotheses of Theorem 20 is often called *inductive*.]

Theorem 20 is named Zorn's Lemma after M. Zorn, who first observed its extreme usefulness. In its most routine applications, X is usually a set partially ordered by $\subset$; for example, in the proof that every commutative ring with unity has a proper maximal ideal, or in the proof that every vector space has a basis. [In current mathematical writing, proofs of the latter type are, regrettably, often reduced to the phrase " by a standard zornification we have . . . ."]

We now make a not-so-routine application of Zorn's Lemma to the proof of a third basic set-theoretic principle, the Well-Ordering Principle of Zermelo. The latter will easily imply the Axiom of Choice, and the chain of implications will be closed.

Definition 2.4

1^0 A POS $(X, \leqslant)$ is *well-ordered* *iff* every non-empty subset A of X contains a lower bound for itself; that is, there exists $e \in A$ such that $e \leqslant a$, for all $a \in A$. Such an e is often called the *least* [or minimum] *element* of A. [Note that every well-ordered $(X, \leqslant)$ is totally ordered, since every subset $\{x, y\}$ has a lower bound for itself.]

2^0 If X is a set, and it is possible to define a partial ordering $\leqslant$ on X such that $(X, \leqslant)$ is well-ordered, then we say that X can be well-ordered.

Example: The system of positive integers in its natural order is well-ordered.

Theorem 21 [*OAXC*] [*Well-ordering Principle*]

Every set can be well-ordered.

Proof outline: Let W be the set of all well-ordered systems $(E, \leqslant)$ such that $E \subset X$. [X is the set to be well-ordered.] Define a partial ordering $|\leqslant|$ on W as follows: $(E_0, \leqslant_0) \ |\leqslant| \ (E_1, \leqslant_1)$ *iff*

1^0 $E_0 \subset E_1$,

2^0 $\leqslant_0$ is the restriction of $\leqslant_1$ to E_0, and

3^0 $x \in (E_1 - E_0)$ implies $x \geqslant y$ for all $y \in E_0$.

Show that $(W, |\leqslant|)$ satisfies the hypotheses of Zorn's Lemma, and look at the maximal element whose existence is implied by that lemma.

Theorem 22

The Well-Ordering Principle implies the Axiom of Choice.

This completes the proof that the Axiom of Choice, Zorn's Lemma, and the Well-Ordering Principle are equivalent.

Note that from now on AXC is to be used freely, and the hypothesis OAXC will no longer appear.

FURTHER PROPERTIES OF WELL-ORDERED SETS

We now make a closer study of the properties and uses of well-ordered sets.

Definition 2.5

Let $(X, \leqslant)$ be a QOS.

1^0 A *segment* of $(X, \leqslant)$ is a subset S of X such that $(y \in S, x \in X, \text{ and } x \leqslant y)$ imply $x \in S$.

2^0 Let $(X', \leqslant)$ also be a QOS. [We are using $\leqslant$ to denote two *different* relations.] An *isomorphism* between $(X, \leqslant)$ and $(X', \leqslant)$ is a 1–1 function f with domain X and range X' such that $x \leqslant y$ implies $f(x) \leqslant f(y)$ for all $x, y \in X$, and such that $x' \leqslant y'$ implies $f^{-1}(x') \leqslant f^{-1}(y')$ for all $x', y' \in X'$. [It should be observed that f^{-1} is then an isomorphism between $(X', \leqslant)$ and $(X, \leqslant)$, so that the verbal symmetry is justified.] $(X, \leqslant)$ and $(X', \leqslant)$ are said to be *isomorphic*.

Theorem 23

Let $(X, \leqslant)$ be well-ordered.

1^0 All unions and intersections of segments of X; and all segments of segments of X, are again segments of X.

2^0 For each $x \in X$, let $S_x = \{y \in X \mid y < x\}$.   Then every S_x is a segment of X, and for each segment S of X, except X itself, there exists an x such that $S = S_x$.

Theorem 24

The set X^* of all segments of a well-ordered $(X, \leqslant)$ is itself well-ordered by inclusion.   The function $f: X \to X^*$ defined by $f(x) = S_x$ is an isomorphism between $(X, \leqslant)$ and $(X^* - \{X\}, \subset)$.

The next theorem is a powerful extension of the familiar principle of finite induction.   [Cf., Theorem 27 and its associated example.]

Theorem 25 [*Principle of Transfinite Induction* 1]

Let $(X, \leqslant)$ be well-ordered, and, for each $x \in X$, let ' $P(x)$ ' be some assertion.   *If* for each $x \in X$, the hypothesis ' $P(y)$ for all $y < x$ ' implies ' $P(x)$ ', *then* $P(x)$, for all $x \in X$.

[The formulation of a principle of this kind presents notational difficulties.   The symbol ' $P(x)$ ' is not the assertion but its abbreviation; however, we are using the single quotes in this case to name the assertion and *not* its abbreviation.   We might have written: let ' ... x ... ' be some assertion. But this is typographically inconvenient.]

Definition 2.6

Let $(X, \leqslant)$ be well-ordered.

1^0 The *successor* of an element $a \in X$ is the least [= minimum] element of the set of all $x \in X$ such that $x > a$. If the latter set is empty, then a has no successor. The successor of a is often denoted by '$a+1$'.

2^0 The *predecessor* of an element $a \in X$ is the greatest [= maximum] element of the set of all $x \in X$ such that $x < a$. If the latter set has no largest element, then a has no predecessor.

Theorem 26

Let $(X, \leqslant)$ be well-ordered.

1^0 Every $a \in X$ has a successor, except the maximal element of X (if there is one).

2^0 If $a \in X$ has a successor $a+1$, then a is the predecessor of $a+1$, and if a has a predecessor b, then $a = b+1$.

Theorem 27 [*Principle of Transfinite Induction* 2]

Let $(X, \leqslant)$ be well-ordered, and, for each $x \in X$, let '$P(x)$' be some assertion. *If* for each $x \in X$ without a predecessor, the hypothesis '$P(y)$ for all $y < x$' implies '$P(x)$'; *and if*, furthermore, '$P(x)$' implies '$P(x+1)$' for each $x \in X$ with a successor, *then* $P(x)$ for all $x \in X$.

Example: If X is the set of strictly positive integers, then Theorem 25 is the so-called weak principle of finite induction. On the other hand, the only strictly positive integer without a predecessor is 1, while all such integers have successors. Thus, Theorem 27 becomes the ordinary principle of finite induction.

We have now reached the point at which essential use must be made of Axiom 5. It implies that the f and the U introduced in the next definition are always sets, even though the G in that definition may not be a set.

Definition 2.7 [*Definition by Transfinite Induction*]

Let $(X, \leqslant)$ be a well-ordered set, and let G be a generalized function. A function f with domain X *is defined* [from G] *by transfinite induction iff* $f \,|\, S_x$ is in the domain of G and $f(x) = G(f \,|\, S_x)$, for all $x \in X$.

A set U *is defined* [from G] *by transfinite induction iff* $U = pr_2 f$ for some function f defined from G by transfinite induction.

*Theorem 28

Let $(X, \leqslant)$ be a well-ordered set, and let Y be a class. Let G be a generalized function whose domain includes the class $p(X \times Y)$ and such that $pr_2 \, G \subset Y$. Then there exists a *unique* function $f : X \to Y$ such that $f(x) = G(f|S_x)$ for all $x \in X$.

We now prepare for the definition of ordinal numbers. The next theorem shows that for any two well-ordered sets, one is just like a part of the other.

Lemma 29.1

Let $(X, \leqslant)$ and $(Y, \leqslant)$ be well-ordered, and let f and g be *increasing* [$x \leqslant x'$ implies $f(x) \leqslant f(x')$ and $g(x) \leqslant g(x')$] mappings of X into Y such that $f[X]$ is a segment of Y and such that g is strictly increasing [$x < x'$ implies $g(x) < g(x')$]. *Then* $f(x) \leqslant g(x)$ for all $x \in X$.

Theorem 29

Let $(X, \leqslant)$ and $(Y, \leqslant)$ be well-ordered. Then *either* there exists an isomorphism of $(X, \leqslant)$ onto a segment of $(Y, \leqslant)$ *or* an isomorphism of $(Y, \leqslant)$ onto a segment of $(X, \leqslant)$. In either case, the isomorphism is unique.

Proof outline: Let E be the set of all mappings of subsets of X into Y, each of which is defined on a segment of X and is an isomorphism of that segment onto a segment of Y. Partially order E as follows: if $f, g \in E$, then $f \leqslant g$ *iff* $(pr_1 g) \supset (pr_1 f)$ and $g \mid (pr_1 f) = f$. [That is, g is an *extension* of f.] Show that $(E, \leqslant)$ is inductive and apply Zorn's Lemma. Use Lemma 29.1 to prove the uniqueness.

COROLLARY 29.1. The only isomorphism of a well-ordered $(X, \leqslant)$ onto a segment of $(X, \leqslant)$ is the identity mapping of X onto itself.

COROLLARY 29.2. Let $(X, \leqslant)$ and $(Y, \leqslant)$ be well-ordered. If there exists an isomorphism f of $(X, \leqslant)$ onto a segment T of $(Y, \leqslant)$ and an isomorphism g of $(Y, \leqslant)$ onto a segment S of $(X, \leqslant)$, then $S = X$, $T = Y$, $f = g^{-1}$, and $g = f^{-1}$.

ORDINAL NUMBERS

Definition 2.8

Let $(X, \leqslant)$ be well-ordered, and let a function f be defined by transfinite induction as follows:

$f(x) = pr_2(f \mid S_x)$ for all $x \in X$. [Here, $G = pr_2$, and Y is the class of all sets, cf. Theorem 28.]

The ordinal number ord $(X, \leqslant)$ of $(X, \leqslant)$ *is defined* to be $pr_2 f$.

A set $\boldsymbol{a}$ is an *ordinal iff* there exists a well-ordered $(X, \leqslant)$ such that $\boldsymbol{a} = $ ord $(X, \leqslant)$.

Examples:

1^0 Let $(X, \leqslant) = (\varnothing, \varnothing)$.   Then ord $(X, \leqslant) = \varnothing$. $[S_x = \varnothing$ for all $x \in X$; thus $f \mid S_x$ and $pr_2(f \mid S_x)$ are empty.]

2^0 Let $X = \{0, 1, 2\}$, with $0 < 1 < 2$.   Then

$$\text{ord } (X, \leqslant) = \Big\{\varnothing, \ \{\varnothing\}, \ \{\varnothing, \{\varnothing\}\}\Big\}.$$

$[S_0 = \varnothing$; thus $f(0) = \varnothing$.

$S_1 = \{0\}$; thus $f(1) = pr_2(f \mid S_1) = \{\varnothing\}$.

$S_2 = \{0, 1\}$;   thus $f(2) = \{\varnothing, \{\varnothing\}\}.]$

Lemma 30.1

Let $(X, \leqslant)$ be well-ordered, and let f be the function of Definition 2.8, so that ord $(X, \leqslant) = pr_2 f$.   Then

1^0 $f(x) \notin f(x)$ for all $x \in X$.

2^0 Let $x \in X$ have a successor $x+1$.   Then

$$f(x+1) = f(x) \cup \{f(x)\}.$$

Theorem 30

Let $(X, \leqslant)$ and f be as in the lemma, and let $x, y \in X$.

1^0 $x \neq y$ iff $f(x) \neq f(y)$.

2^0 $x \leqslant y$ iff $f(x) \subset f(y)$.

3^0 $x < y$ iff $f(x) \in f(y)$.

COROLLARY 30.1.   On ord $(X, \leqslant)$, let $|\leqslant|$ be the relation defined by $a \mid \leqslant \mid b$ iff $(a \in b$ or $a = b)$.   Then $(\text{ord } (X, \leqslant), |\leqslant|)$ and $(\text{ord } (X, \leqslant), \subset)$ are both well-ordered, and both are isomorphic to $(X, \leqslant)$.

COROLLARY 30.2. Every member of, and every segment of, (ord $(X, \leqslant)$, $\subset$) is again an ordinal.

Theorem 31

Let $(X, \leqslant)$ and $(Y, \leqslant)$ be well-ordered. Then $(X, \leqslant)$ and $(Y, \leqslant)$ are isomorphic iff ord $(X, \leqslant)$ = ord $(Y, \leqslant)$.

Theorem 32.1

Let $(X, \leqslant)$ and $(Y, \leqslant)$ be well-ordered. Then either

$$\text{ord}\,(X, \leqslant) \subset \text{ord}\,(Y, \leqslant)$$

or vice versa.

Theorem 32.2

Let $(X, \leqslant)$ and $(Y, \leqslant)$ be well-ordered. Then exactly one of the following is true:

1^0 ord $(X, \leqslant) \in$ ord $(Y, \leqslant)$.

2^0 ord $(X, \leqslant) =$ ord $(Y, \leqslant)$.

3^0 ord $(Y, \leqslant) \in$ ord $(X, \leqslant)$.

COROLLARY 32.1.   Unequal ordinals are not isomorphic.

COROLLARY 32.2.   Every well-ordered $(X, \leqslant)$ is isomorphic to one and only one ordinal.

The next theorem enables us to define a successor for any ordinal.

Theorem 33

Let $(X, \leqslant)$ be well-ordered, and suppose $m \notin X$. Then $(X \cup \{m\}, \leqslant)$ is well-ordered, if $\leqslant$ is extended to $X \cup \{m\}$ so that $x \leqslant m$ for all $x \in (X \cup \{m\})$. Furthermore,

$$\operatorname{ord}(X \cup \{m\}, \leqslant) = \operatorname{ord}(X, \leqslant) \cup \{\operatorname{ord}(X, \leqslant)\}.$$

Definition 2.9

1^0 If a is an ordinal, then the ordinal $a \cup \{a\}$ is called the *successor* of a and is denoted by '$a+1$'. As in Definition 2.6 [cf., Theorem 26], we call an ordinal b the *predecessor* of a iff $a = b+1$.

2^0 An *ordinal* $a \neq 0$ is a *limit ordinal* iff a has no predecessor. [For example, ω defined below is a limit ordinal.] Here and subsequently $0 = \emptyset$, the zero ordinal.

Theorem 34

Let a and b be ordinals.

1^0 If $a \neq \emptyset$, then *either* a is a limit ordinal and $a = \cup a$ *or else* a is the successor of $\cup a$.

2^0 If $a \neq b$, then either $a \in b$ or $b \in a$.

3^0 If $a \in b$, then a is a segment of $(b, \subset)$. And if b^* is the set of all segments of $(b, \subset)$ except b itself, then $b = b^*$.

If a and b are ordinals, we shall frequently write $a \leqslant b$ and $a < b$ instead of $a \subset b$ and $a \in b$, respectively. This usage is justified as follows: for any ordinals a and b, either $a \subset b$ or $b \subset a$, and if $a \neq b$, then either $a \in b$ or $b \in a$. Suppose that $a \subset b$. Then both a and b are members of $b+1$. Now $(b+1, \subset)$ is well-ordered; so if we write $\leqslant$ instead of $\subset$, then $a \leqslant b$ is simply a statement about order in $(b+1,\ \subset)$. On the other hand, if $a \neq b$, then $a \subset b$ implies $a \in b$; that is, $a \leqslant b$ and $a \neq b$ imply $a < b$; thus $a < b$ is also simply a statement about order in $(a+1,\ \subset)$. However, we must resist, in order to avoid the Burali–Forti paradox[2], any temptation to say that the class of *all* ordinals is a set well-ordered by $\leqslant$. In fact, the paradox merely proves that it is a proper class.

[2] This paradox consists of proving that there is a greatest ordinal, thus contradicting the fact that every ordinal has a successor greater than it.

Chapter *III*

Ordinal and cardinal arithmetic

We are now in a position to define the arithmetic operations of addition and multiplication for ordinals.

Definition 3.1

Let A and B be sets, and let i and j be such that both i and $j \notin A \cup B$ and $i \neq j$. Then the set $(A \times \{i\}) \cup (B \times \{j\})$ is called a *disjoint union* of A and B. Evidently,

$$(A \times \{i\}) \cap (B \times \{j\}) = \varnothing.$$

[A disjoint union of A and B is sometimes called a *direct sum* (or union) of A and B; similarly, $A \times B$ is sometimes called their *direct product*.]

Theorem 35

Let $(X, \leqslant)$ and $(Y, \leqslant)$ be well-ordered.

1^0 Let $Z = (X \times \{i\}) \cup (Y \times \{j\})$ be a disjoint union of X and Y. Then Z is well-ordered by the following relation $\leqslant$:

(a) If $z, z' \in Z$ are both in $(X \times \{i\})$ or both in $(Y \times \{j\})$, then $z \leqslant z'$ iff $(pr_1 z) \leqslant (pr_1 z')$.

(b) If $z \in (X \times \{i\})$ and $z' \in (Y \times \{j\})$, then $z < z'$.

2^0 Let $Z = X \times Y$. Then Z is well-ordered by the following relation $\leqslant$:

(a) If $z, z' \in Z$, and if $(pr_1 z) < (pr_1 z')$, then $z < z'$.

(b) If $z, z' \in Z$, and if $(pr_1 z) = (pr_1 z')$, then $z \leqslant z'$ iff $(pr_2 z) \leqslant (pr_2 z')$.

[This is called the *lexicographic* ordering of $X \times Y$.]

3^0 If $(X, \leqslant)$ and $(X', \leqslant)$ are isomorphic, if $(Y, \leqslant)$ and $(Y', \leqslant)$ are isomorphic, and if $(Z', \leqslant)$ is defined as in 1^0 [or as in 2^0], then $(Z, \leqslant)$ and $(Z', \leqslant)$ are isomorphic. [Here Z is defined relative to X and Y, Z' relative to X' and Y'.]

Theorem 35 permits us to make the following definition.

Definition 3.2

Let $(X, \leqslant)$ and $(Y, \leqslant)$ be well-ordered, and let

$$a = \text{ord}(X, \leqslant) \quad \text{and} \quad b = \text{ord}(Y, \leqslant).$$

1^0 Let $(Z, \leqslant)$ be the well-ordered direct sum defined in Theorem 35.1^0. Then $a + b = \text{ord}(Z, \leqslant)$ by definition.

2^0 Let $(Z, \leqslant)$ be the well-ordered direct product defined in Theorem 35.2^0. Then $ba = \text{ord}(Z, \leqslant)$ by definition. *Note* the order. [We shall define the power a^b in Definition 3.9.]

We shall now introduce an axiom which guarantees the existence of many ordinals and in particular of the natural numbers. We could simply assume that the natural numbers $\{0, 1, 2, \ldots\}$ are already known to exist and to form a well-ordered set. Then we could define $0 = \text{ord}(\emptyset, \subset)$; and for $n > 0$,

$n = \text{ord}(\{0, ..., n-1\}, \leqslant)$; and finally $\omega = \text{ord}(\{0, 1, ...\}, \leqslant)$. Of course, this leaves the composition of the sets 0, 1, etc. very much up in the air.

It is axiomatically more satisfactory, and not much harder, to proceed differently by *defining* the natural numbers. We then *prove* the Peano postulates and finally adjoin the class of natural numbers to our sets by another axiom.

Definition 3.3

1^0 A *natural number* is an ordinal a such that $(a, \supset)$ is well-ordered.

[Since a is an ordinal, we know that $(a, \subset)$ is well-ordered. The definition asserts that $(a, \leqslant)$ is well-ordered where for $x, y \in a$, we have $x \leqslant y$ iff $x \supset y$. This is equivalent to the statement that every non-empty subset of a has a maximum element with respect to $(a, \subset)$.]

2^0 ω is by definition the class of all natural numbers.

AXIOM 6 ω is a set.

The first five parts of the following theorem, which do not use Axiom 6, are the Peano postulates for the natural numbers, starting with 0 [$= \varnothing$] rather than with 1 [$= 0 + 1$].

Theorem 36

1^0 $\varnothing \in \omega$.

2^0 If $a \in \omega$, then $a + 1 \in \omega$.

3^0 If $a \in \omega$, then $a + 1 \neq \varnothing$.

4^0 If $a, b \in \omega$ and $a + 1 = b + 1$, then $a = b$.

5^0 Let $X \subset \omega$. If $\varnothing \in X$, and if $\boldsymbol{a}+1 \in X$ whenever $\boldsymbol{a} \in X$, then $X = \omega$.

6^0 The set $(\omega, \subset)$ is well-ordered; in fact, ord $(\omega, \subset) = \omega$.

We observe that on the basis of Axiom 6 we can define the set of real numbers, using standard constructions: An equivalence relation on $\omega \times \omega$ gives us the set of positive rationals Q_+; another equivalence on $Q_+ \times Q_+$ gives us the set of all the rationals Q; and a final relation in $p(Q)$ produces the set of real numbers.

In the next few theorems we list some properties of ordinal arithmetic, most of which resemble corresponding properties of natural numbers; however, some properties of the operations on natural numbers are not possessed by the general ordinal operations. For example, $\omega + \boldsymbol{1} = \omega + 1$, but $\boldsymbol{1} + \omega = \omega$. Furthermore, if $\boldsymbol{2} = \boldsymbol{1} + 1$ $[= \boldsymbol{1} + \boldsymbol{1}]$, then $\omega\boldsymbol{2} = \omega + \omega$, but $\boldsymbol{2}\omega = \omega \neq \omega\boldsymbol{2}$, and so, since $\boldsymbol{1} + \boldsymbol{1} = \boldsymbol{2}$, we also have $(\boldsymbol{1} + \boldsymbol{1})\,\omega \neq \boldsymbol{1}\omega + \boldsymbol{1}\omega$. Incidentally, for later use, let $\boldsymbol{3} = \boldsymbol{2} + 1$ and $\boldsymbol{4} = \boldsymbol{3} + 1$.

Theorem 37

Let $\boldsymbol{a}$, $\boldsymbol{b}$, and $\boldsymbol{c}$ be ordinals. Then

1^0 $(\boldsymbol{a} + \boldsymbol{b}) + \boldsymbol{c} = \boldsymbol{a} + (\boldsymbol{b} + \boldsymbol{c})$.

2^0 $(\boldsymbol{a}\boldsymbol{b})\,\boldsymbol{c} = \boldsymbol{a}(\boldsymbol{b}\boldsymbol{c})$.

3^0 $\boldsymbol{a}(\boldsymbol{b} + \boldsymbol{c}) = \boldsymbol{a}\boldsymbol{b} + \boldsymbol{a}\boldsymbol{c}$.

Lemma 38.1

Let $(X, \leqslant)$ be well-ordered, and let $\{x_n,\ n \in \omega\}$ be a sequence of elements of X such that $x_{n+1} \leqslant x_n$ for each $n \in \omega$. Then there exists $p \in \omega$ such that $x_q = x_p$ for all $q \leqslant p$.

Lemma 38.2

Let $(X, \leqslant)$ be well-ordered, and suppose that $Y \subset X$. Then ord $(Y, \leqslant) \leqslant$ ord $(X, \leqslant)$.

[The second lemma is not trivial, since a *subset* of an ordinal need *not* be ordinal.]

*Theorem 38

Let a, b, and c be any ordinals, and let d be an ordinal such that $0 < d$. Then

1^0 $a+1 = a+1$; $a+0 = a = 0+a$; $a0 = 0 = 0a$.

2^0 $a < b$ iff $a+1 \leqslant b$.

3^0 If $a < b$, then $c+a < c+b$, $a+c \leqslant b+c$, $ac \leqslant bc$, and $da < db$.

4^0 Each of the following conditions implies that $a < b$: $c+a < c+b$, $a+c < b+c$, $da < db$, and $ad < bd$.

5^0 Each of the following conditions implies that $a = b$: $c+a = c+b$, and $da = db$.

6^0 $a \leqslant b$ iff there exists c such that $b = a+c$. Such a c is unique, and is $\leqslant b$. [c is sometimes denoted by $(-a)+b$.]

7^0 If $c < ab$, then there exist ordinals u and v such that $c = av+u$, where $u < a$ and $v < b$. Such u and v are unique.

CARDINAL NUMBERS

After a preliminary theorem, we shall define the *cardinal* numbers.

Theorem 39

Every nonempty set C of ordinals has a least element; that is, there exists $a \in C$ such that $a \leqslant c$ for all $c \in C$.

Definition 3.4

Let X be a set, and let $C = \{c \mid c = \text{ord}(X, \leqslant)$ for *some* well-ordering $\leqslant$ of $X\}$. By Theorem 39, C has a least element a. We call a the *cardinal number* of X and write $a = \text{card } X$.

Definition 3.5

Two sets X and Y are *equipotent* [or have the same cardinality] *iff* there exists a 1–1 function of X *onto* Y. [And hence also a function of Y onto X.]

Theorem 40

Let X and Y be sets. Then

1^0 X and Y are equipotent iff card $X = $ card Y.

2^0 X is equipotent with a subset of Y iff card $X \leqslant Y$.

3^0 X is equipotent with a subset of Y, but Y is equipotent with *no* subset of X, iff card $X < $ card Y.

COROLLARY 40.1 Let $A \subset X$ and $B \subset Y$. If X and B are equipotent, and if Y and A are equipotent, then X and Y are equipotent.

This corollary and some of the subsequent theorems can be proved without using ordinal number theory. In fact, much of our development of cardinal number theory could be carried out independently of ordinal numbers. If both theories are to be developed, however, it is more elegant to discuss ordinals first, and then use them in discussing cardinals. Otherwise, each theory requires a separate development.

We next define the arithmetic operations for cardinal numbers. These are *not* the same as the operations already defined for ordinals. Thus, when A and B are cardinals [and hence also ordinals in our definition], the sum $A+B$, etc. [as ordinals] is *in general* different from the sum $A+B$, etc. [as cardinals]. If A and B are specifically assumed to be cardinals, then $A+B$, etc. will denote the cardinal operations, unless otherwise stated.

Definition 3.6

Let A and B be cardinals.

1^0 $A+B = \operatorname{card} Z$, where Z is a disjoint union of A and B.

2^0 $AB = \operatorname{card}(A \times B)$.

3^0 $A^B = \operatorname{card} Z$, where $Z = \{f \colon B \to A\}$.

Theorem 41

Let X and Y be sets such that $X \cap Y = \varnothing$, and let $A = \operatorname{card} X$ and $B = \operatorname{card} Y$. Then

$$A+B = \operatorname{card}(X \cup Y),$$

$$AB = \operatorname{card}(X \times Y),$$

and $$A^B = \operatorname{card}\{f \colon Y \to X\}.$$

Theorem 42

Let A, B, and C be cardinals.   Then

1^0 $A+B = B+A$, and $AB = BA$.

2^0 $A+(B+C) = (A+B)+C$, and $A(BC) = (AB)C$.

3^0 $A(B+C) = AB+AC$.

Theorem 43

The elements of ω are cardinals, and for them the cardinal and ordinal operations of addition and multiplication coincide. Furthermore, if A and B are any cardinals, then $A+0 = A1 = A$, $A+1 = B+1$ implies $A = B$, and $A0 = 0$.

We shall now give precise definitions of the terms *finite* and *infinite*. This can be done in several different ways, depending partly on taste and partly on the particular system of set theory being developed.

Lemma

The ordinal ω of the natural numbers is a cardinal.

Definition 3.7

1^0 A set X is *finite iff* card $X < \omega$. A set X is *infinite iff* it is *not* finite.

2^0 A set X is *denumerable* [or countable] *iff* card $X \leqslant \omega$. It is *denumerably infinite iff* card $X = \omega$.

3^0 A set X is *nondenumerable* [or uncountable] *iff* card $X > \omega$.

Theorem 44

The following statements about a set X are equivalent:

1^0 X is infinite.

2^0 (card X) $\geqslant \omega$.

3^0 X has the same cardinality as a *proper* subset of itself.

4^0 (card X)$+ \mathbf{1} =$ (card X).   [Cardinal sum.]

Lemma

For all cardinals A, card $A = A$.

Definition 3.8

A cardinal A is *finite iff* $A < \omega$; otherwise A is *infinite*— or transfinite.   [The lemma shows that Definitions 3.7 and 3.8 are consistent.]

We now give the definition of a^b and some of its properties.

EXPONENTIATION

*Theorem 45

1^0 Let $(X, \leqslant)$ and $Y, \leqslant)$ be well-ordered, and let x_0 be the least element of $(X, \leqslant)$. Let $G = \{g : Y \to X$ such that $g(y) = x_0$ for all $y \in Y$ except, possibly, for *finitely* many y.} Then G is well-ordered by the following relation $\leqslant$: For $g, h \in G$, let $A = \{y \in Y \mid g(y) \neq h(y)\}$, and let a be the *largest* element of A.   [A is always finite, and if $A = \varnothing$, then $g = h$.]   Then $g \leqslant h$ iff $g(a) \leqslant g(a)$.

2^0 If $(X, \leqslant)$ and $X', \leqslant)$ are isomorphic, if $(Y, \leqslant)$ and $(Y', \leqslant)$ are isomorphic, and if $(G', \leqslant)$ is defined as in 1^0 using X' and Y', then $(G, \leqslant)$ and $(G', \leqslant)$ are isomorphic.

Definition 3.9

Let $(X, \leqslant)$ and $(Y, \leqslant)$ be well-ordered, and let

$$a = \text{ord}\,(X, \leqslant) \quad \text{and} \quad b = \text{ord}\,(Y, \leqslant).$$

Let $(G, \leqslant)$ be the set defined in Theorem 45.1. Then a^b is ord $(G, \leqslant)$ by definition.

Theorem 46

If a, b, and c are ordinals, then

$$a^b . a^c = a^{b+c}, \quad \text{and} \quad (a^b)^c = a^{bc}.$$

Referring back to Theorem 43, observe that in ω the cardinal and ordinal operations of exponentiation coincide, and that for any cardinal A one has $A^1 = A$, $1^A = 1$, $A^0 = 1$; and if $A \neq 0$, then $0^A = 0$. The definition of exponentiation for ordinals had to be delayed until the term " finite " was defined.

Chapter IV

Further properties of ordinal and cardinal numbers

We can extend the cardinal operations to infinite sums and products. To that end we make a natural extension of Definition 3.1. The index-class I is *always* assumed to be a *set*.

Definition 4.1

Let $\{X_i,\ i \in I\}$ be a family of sets. Then the set

$$\cup\{X_i \times \{i\},\ i \in I\}$$

is called a *disjoint union* of the family. [Note that

$$(X_i \times \{i\}) \cap (X_j \times \{j\}) = \varnothing,$$

if $i \neq j$.]

Definition 4.2

Let $\{A_i,\ i \in I\}$ be a family of cardinals.

1^0 $S\{A_i,\ i \in I\} = \text{card} \cup \{A_i \times \{i\},\ i \in I\}$ by   definition. [That is, the sum is the cardinal of a disjoint union of the A_i.]

2^0 $P\{A_i, \ i \in I\} = \text{card } \Pi\{A_i, \ i \in I\}$ by definition.

[We write P for the cardinal product to avoid confusion with the Cartesian product Π.]

Observe that if $\{X_i, \ i \in I\}$ is any family such that card $X_i = A_i$, and if Z is a disjoint union of $\{X_i, \ i \in I\}$, then

$$S\{A_i, \ i \in I\} = \text{card } Z$$

and $$P\{A_i, \ i \in I\} = \text{card } \Pi\{X_i, \ i \in I\}.$$

We can now prove a more general form of Theorem 42.

Lemma 47.1

Let $\{A_i, \ i \in I\}$ be a family of cardinals, and let f be a 1–1 mapping of a set K onto I. Then

$$S\{A_{f(k)}, \ k \in K\} = S\{A_i, \ i \in I\}$$

and $$P\{A_{f(k)}, \ k \in K\} = P\{A_i, \ i \in I\}.$$

Theorem 47

Let $\{A_i, \ i \in I\}$ be a family of cardinals, and let $\{J_l, \ l \in L\}$ be a disjoint family of subsets of I such that $\cup\{J_l, \ l \in L\} = I$. [Such a family is called a *partition* of I.]   Then

$$S\{A_i, \ i \in I\} = S\{S\{A_i, \ i \in J_l\}, \ l \in L\},$$

and $$P\{A_i, \ i \in I\} = P\{P\{A_i, \ i \in J_l\}, \ l \in L\}.$$

Let $F = \Pi\{J_i, \ i \in I\}$ where $\{J_i, \ i \in I\}$ is any indexed set; then

$$P\{S\{A_{ij}, \ j \in J_i\}, \ i \in I\} = S\{P\{A_{if(i)}, \ i \in I\}, \ f \in F\}$$

where $\{A_{ij}, \ j \in J_i, \ i \in I\}$ is a family of cardinals.

Lemma 48.1

Let A and B be cardinals, and let I be a set such that

$$B = \operatorname{card} I.$$

For each $i \in I$, let $A_i = A$. Then

$$A^B = P\{A_i, \ i \in I\},$$

and

$$A \cdot B = S\{A_i, \ i \in I\}.$$

Theorem 48

Let A, B, and C be cardinals, and let $\{A_i, \ i \in I\}$ and $\{B_i, \ i \in I\}$ be families of cardinals. Then

1^0 $A^{S\{B_i, \ i \in I\}} = P\{A^{B_i}, i \in I\}.$

2^0 $(P\{A_i, \ i \in I\})^B = P\{A_i{}^B, \ i \in I\}.$

3^0 $A^{BC} = (A^B)^C.$

Theorem 49

Let $\{A_i, \ i \in I\}$ and $\{B_i, \ i \in I\}$ be families of cardinals such that $A_i \geqslant B_i$ for all $i \in I$. Then

$$S\{A_i, \ i \in I\} \geqslant S\{B_i, \ i \in I\},$$

and

$$P\{A_i, \ i \in I\} \geqslant P\{B_i, \ i \in I\}.$$

COROLLARY 49.1. If A, A', B, B' are cardinals such that $A \leqslant A'$, $B \leqslant B'$, and $A' > 0$, then $A^B \leqslant A'^{B'}$.

*Theorem 50

Let $\{A_i, \ i \in I\}$ and $\{B_i, \ i \in I\}$ be families of cardinals and suppose $B_i \geqslant 2$ for all $i \in I$.

1^0 If $A_i \leqslant B_i$ for all $i \in I$, then $S\{A_i, \ i \in I\} \leqslant P\{B_i, \ i \in I\}.$

2^0 If $A_i < B_i$ for all $i \in I$, then $S\{A_i, \ i \in I\} < P\{B_i, \ i \in I\}.$

Theorem 51

If A is the cardinal of a set X, then 2^A is the cardinal of the set of all subsets of X. Furthermore, if A is any cardinal, then $A < 2^A$.

*Theorem 52

If A is an infinite cardinal, then $A^2 = A$.

Proof outline: Let $A = \text{card } X$, for some set X. Show that there exists $D \subset X$ such that $\text{card } D = \omega$; show further that $\text{card } (D \times D) = \text{card } D$. Now let M be the set of all pairs (U, f) such that $D \subset U \subset X$ and such that f is a 1–1 mapping of U onto $U \times U$; define $(U, f) \leqslant (V, g)$ iff $U \subset V$ and $f = g \mid U$. Then M is inductive. If (F, f) is a maximal element, prove that $\text{card } F = \text{card } X$.

COROLLARY 52.1. If A is an infinite cardinal, then $A^n = A$, for all integers $n \geqslant 1$.

COROLLARY 52.2. Let $\{A_i,\ i \in I\}$ be a family of cardinals.

   1^0 If I is finite, if $A_i \neq 0$ for all $i \in I$, and if the largest cardinal A_k in the family is infinite, then $P\{A_i,\ i \in I\} = A_k$.

   2^0 If A is an infinite cardinal such that $\text{card } I \leqslant A$, and if $A_i \leqslant A$ for all $i \in I$, then $S\{A_i,\ i \in I\} \leqslant A$. In this case, if at least one A_i equals A, then $S\{A_i,\ i \in I\} = A$.

COROLLARY 52.3. Let A and B be cardinals $\neq 0$, at least one of which is infinite. Then

$$AB = A + B = \max\,[A, B].$$

Theorem 53

Let f be a mapping of a set X *onto* an infinite set Y such that for each $y \in Y$, $f^{-1}(y)$ is countable. Then card X = card Y.

Theorem 54

The set F of all *finite* subsets of an infinite set X has the same cardinality as X.

*Theorem 55

Let X and Y be infinite sets such that card $Y \leqslant$ card X. Then the set of all mappings of X onto Y, the set of all mappings of X into Y, the set of all mappings of subsets of X into Y, and the set of all subsets of X have the same cardinality.

COROLLARY 55.1. If $A \leqslant B$, then $A^B = 2^B$, where B is infinite and $A \geqslant 2$.

THE ALEPHS

For any well-ordered set $(X, \leqslant)$ we have constructed a standard well-ordered set $\left(\text{ord}\,(X, \leqslant),\ \subset\right)$ isomorphic to it. Now since a set of cardinals is in particular a set of ordinals, such a set is well-ordered and has an ordinal number. This kind of application of ordinals to cardinals will lead us to the definition of the " alephs ", which are simply an ordinal arrangement of the *infinite* cardinals.

Definition 4.3

Let A be an infinite cardinal, and let C_A be the set of all infinite cardinals C such that $C < A$. The *aleph index* $i(A)$ of the cardinal A is *by definition* ord $(C_A, \subset)$.

Theorem 56

Let A and B be infinite cardinals.

1^0 $A = B$ iff $i(A) = i(B)$.

2^0 $A \leqslant B$ iff $i(A) \leqslant i(B)$.

3^0 $A < B$ iff $i(A) < i(B)$.

COROLLARY 56.1.   Let $(M, \leqslant)$ be a set of infinite cardinals in their natural order, and let $N = \{i(A) \mid A \in M\}$.   Then $(M, \leqslant)$ and $(N, \leqslant)$ are isomorphic.

Theorem 57

For each ordinal a there exists a unique infinite cardinal A such that $a = i(A)$.

COROLLARY 57.1.   Let A be an infinite cardinal.   Then

$$\{i(C) \mid C < A\}$$

is the segment $S_{i(A)}$ of the ordinal numbers.

Definition 4.4

For each ordinal a, the unique infinite cardinal A of Theorem 57 such that $a = i(A)$ is denoted by $\aleph_a$, [or by ω_a, if ordinality is to be emphasized.   Those ordinals that are cardinals are often called *initial* ordinals.]

COROLLARY 57.2.   If a is an ordinal and A is a cardinal such that $\aleph_a \leqslant A \leqslant \aleph_{a+1}$, then $A = \aleph_a$ or $A = \aleph_{a+1}$.

Example: $\aleph_0 = \omega$ [$= \omega_0$].

*Theorem 58

For all ordinals a and all cardinals $A > 0$,

$$\aleph_{a+1}^A = \aleph_a^A \cdot \aleph_{a+1}.$$

Like the Axiom of Choice, the following hypothesis is occasionally needed for certain proofs. It is relatively consistent with the other axioms of set theory in the same sense that AXC is itself relatively consistent; in fact, it is independent of the other axioms as is AXC itself.

GENERALIZED CONTINUUM HYPOTHESIS [GCH]

$$2^{\aleph_a} = \aleph_{a+1}, \quad \text{for all } a.$$

The [classical] continuum hypothesis is the special case of GCH in which $a = 0$. The construction of the real numbers R from the rationals shows that card $R = 2^{\aleph_0}$. Thus the continuum hypothesis asserts that the cardinality of the continuum (that is, of R) is the next largest cardinal after $\aleph_0$: card $R = \aleph_1$.

COROLLARY 57.3 [OGCH]. Let P be the set of all subsets of a set X. If A is a cardinal such that card $X \leqslant A \leqslant$ card P, then either $A =$ card X or $A =$ card P.

COROLLARY 58.1 [OGCH].

$$\aleph_{a+1}^{\aleph_b} = \begin{cases} \aleph_{a+1}, & b \leqslant a \\ \aleph_{b+1}, & b > a \end{cases}$$

The next definition and Theorems 62 and 63 will be formulated for an arbitrary, totally ordered QOS.

Definition 4.5

Let $(X, \leqslant)$ be a totally ordered QOS. A subset E of X is *cofinal* with X *iff* for each $x \in X$ there exists $e \in X$ such that $e \geqslant x$.

Theorem 59

Let E [$\neq \emptyset$] be a set of ordinal numbers. Then $\cap E$ and $\cup E$ are again ordinals. More precisely:

1^0 $\cap E$ is the least element of $(E, \subset)$.

2^0 If $\text{ord}\,(E, \subset)$ has a predecessor, then $\cup E$ is the greatest element of $(E, \subset)$.

3^0 If $\text{ord}\,(E, \subset)$ is a limit ordinal, then $\cup E > c$ for all $c \in E$, and if, furthermore, $a > c$ for all $c \in E$, then $a \geqslant \cup E$. Also, if $F \subset E$, then F is cofinal with E iff $\cup F = \cup E$.

For sets of ordinals E, $\cap E$ may be denoted by $\lim \inf E$, and $\cup E$ by $\lim \sup E$ or $\lim E$, the last notation often being reserved for the situation in Theorem 59.3^0. Of course, with a different development of ordinals, Theorem 59.1^0, 2^0, 3^0 would become definitions, since $\cup E$ and $\cap E$ would not then be ordinals.

The following formulation is equivalent: Let b be a limit ordinal, and let $E = \{a_c,\ c \in b\}$ be a family of ordinals such that if $c, d \in b$ and $c < d$, then $a_c < a_d$. By definition, $\lim\limits_{c<b} a_c$ is the least ordinal e greater than all the a_c. In fact, with our definition of ordinal, $e = \cup E$ and $b = \text{ord}\,(E, \subset)$.

Theorem 60

Let E be a set of ordinals, let $e = \cup E$, and let

$$G = \{\omega_a,\ a \in E\} = \{\aleph_a,\ a \in E\}.$$

Then $\omega_e = \cup G$, and $\aleph_e = SG$.

CLASSIFICATION OF ORDINALS[1]

We have already distinguished various kinds of ordinals: ordinals with predecessors and limit ordinals. The limit ordinals may or may not be initial ordinals [that is, cardinals], but all infinite initial ordinals are limit ordinals. We shall now further refine the classification.

Definition 4.6

An initial ordinal A [= cardinal] is called *critical iff* $i(A) = A$. [Observe that if $i(A)$ has a predecessor, then A is not critical.]

[Note: In a set theory which develops ordinals and cardinals separately, the initial ordinals are *not* cardinals, the initial ordinal corresponding to a given cardinal being the least ordinal whose cardinality equals that cardinal. That such a correspondence always exists is equivalent to AXC, upon which our particular development heavily depends.]

Theorem 61

There exist critical ordinals.

The next theorem prepares the way for Definition 4.7.

Theorem 62

Let $(X, \leqslant)$ be a totally ordered QOS. Then $(X, \leqslant)$ has a *well-ordered* cofinal subset.

Definition 4.7

1^0 Let $(X, \leqslant)$ be a totally ordered QOS, and let $L(X, \leqslant)$ be the set of all cofinal well-ordered subsets of $(X, \leqslant)$. Let

$$L^*(X, \leqslant) = \{\text{ord}\,(E, \leqslant) \mid (E, \leqslant) \in L(X, \leqslant)\}.$$

[1] Many of the remaining theorems are quite difficult and deserve an asterisk, in accordance with the Preface.

Then *by definition* the *final character* $fc(X, \leqslant)$ of $(X, \leqslant)$ is the least element of $L^*(X, \leqslant)$.

2^0 If a is an ordinal, then $fc(a) = fc(a, \subset)$, by definition.

3^0 An ordinal a is *regular iff* $fc(a) = a$; otherwise—that is, $fc(a) < a$—it is *singular*.

[Since cardinals are ordinals, this definition applies in particular to cardinals.]

Theorem 63

Let $(X, \leqslant)$ be a totally ordered QOS.   Then

$$fc(X, \leqslant) \leqslant \text{card } X;$$

furthermore, $fc(X, \leqslant)$ is a regular cardinal.

Theorem 64

Let a be an infinite ordinal.

1^0 If a has a predecessor, then a is singular.

2^0 If a is a limit ordinal, then a is singular iff there exists $E \subset a$ such that ord $(E, \subset) < a$ and $\cup E = a$.

COROLLARY 64.1.   A limit ordinal a is regular iff for every $E \subset a$ such that ord $(E, \leqslant) < a$ it follows that $\cup E < a$.

Theorem 65

1^0 The ordinals **0, 1,** and ω are regular.

2^0 If an ordinal $a \neq \mathbf{1}$ has a predecessor, then it is singular. In fact, $fc(a) = \mathbf{1}$.

3^0 If an ordinal *a* is not a cardinal [that is, not finite and not an initial ordinal], then it is singular.

4^0 All the finite cardinals except **0** and **1** are singular.

5^0 Every infinite cardinal A for which $i(A)$ has a predecessor is regular.

6^0 Every finite cardinal $A \neq \aleph_0$ such that $i(A)$ has no predecessor but such that $i(A) < A$ [that is, not critical] is singular. In fact, $fc(A) = fc(i(A))$, if $i(A)$ has no predecessor.

7^0 There exist singular critical cardinals, that is, cardinals A such that $i(A) = A$ and $fc(A) < A$.

COROLLARY 65.1 [OGCH]. Let *a* be a limit ordinal. Then

$$\aleph_a^{\aleph_b} = \begin{cases} \aleph_a, & b < i(fc\ a) \\ \aleph_{a+1}, & i(fc\ a) \leqslant b < a \\ \aleph_{b+1}, & b \leqslant a \end{cases}$$

COROLLARY 65.2 [OGCH]. $\aleph_a^{\aleph_b} = \aleph_a$ for all $\aleph_b < \aleph_a$ iff *a* has a predecessor or $\aleph_a$ is inaccessible. [See Definition 4.8.]

Examples: [OGCH].

$$\aleph_\omega^{\aleph_0} = \aleph_\omega^{\aleph_1} = \aleph_{\omega+1}; \quad \aleph_{\omega+2}^{\aleph_\omega} = \aleph_{\omega+2}^{\aleph_\omega+1} = \aleph_{\omega+2};$$

$$\aleph_{\omega_1}^{\aleph_0} = \aleph_{\omega_1}; \quad \aleph_{\omega_1}^{\aleph_1} = \aleph_{\omega_1+1}; \quad \aleph_{\omega_\omega+1}^{\aleph_\omega} = \aleph_{\omega_\omega+1};$$

$$\aleph_{\omega_\omega+1}^{\aleph_\omega+1} = \aleph_{\omega_\omega+1+1}; \quad \aleph_{\omega_\omega}^{\aleph_0} = \aleph_{\omega_\omega}^{\aleph_1} = \aleph_{\omega_\omega+1}.$$

The following outline summarizes the classification of ordinals.

I. *a not limit ordinal*

   (A) $fc(a) = 1$, if $a \neq 0$ [all singular except **1**]
   (B) $fc(0) = 0$ [**0** regular]

II. ***a** limit ordinal*

    (A) ***a** not initial*   [all singular]

    (B) *A initial*   [$A = a$]

        1^0 $i(A)$ not limit ordinal   [all regular]

        2^0 $i(A)$ limit ordinal

            (a) *A* not critical   [all singular]

            (b) *A* critical

                (a*) *A* singular

                (b*) *A* regular   [inaccessible] [?]

Recall that

a	initial	iff $a = \text{card } a$
a	not initial	iff $a > \text{card } a$
A	critical	iff $i(A) = A$
A	not critical	iff $i(A) < A$
a	regular	iff $fc(a) = a$
a	singular	iff $fc(a) < a$.

Definition 4.8

    An infinite cardinal *A* is *inaccessible* iff *A* is both regular and critical. [Equivalently: *A* is regular and $i(A) \neq 0$ has no predecessor.]

This is II.B.2.b.b* in the outline. Even assuming both AXC and GCH, one does not know whether any inaccessible cardinals exist! Nor is it known whether the assumption of their existence is relatively consistent with the rest of set theory in the sense that AXC and GCH are relatively consistent. We shall formulate below a hypothesis which implies the existence of inaccessible cardinals. But first we shall see [Theorem 68] what GCH implies for such cardinals.

Theorem 66 [*OGCH*]

Let A be an infinite cardinal. Then the following statements are equivalent:

1^0 A is regular.

2^0 If $\{A_i,\ i \in I\}$ is any family of cardinals such that $\operatorname{card} I < A$ and such that $A_i < A$ for all $i \in I$, then $S\{A_i,\ i \in I\} < A$.

3^0 For all cardinals $B > 0$ it follows that

$$A^B = A \cdot S\{C^B,\ C \in A\}.$$

Theorem 67

The following statement is equivalent to GCH: For each regular cardinal A and each cardinal B such that $0 < B < A$, it follows that $A^B = A$.

Definition 4.9

1^0 An infinite cardinal A is *dominant iff* for all $B, C < A$ one has $B^C < A$.

2^0 An infinite cardinal is *strongly inaccessible iff* it is both inaccessible and dominant.

Theorem 68 [*OGCH*]

Every inaccessible cardinal is strongly inaccessible.

The notion of inaccessibility can be formulated more generally. Let $e \langle C \rangle$ be an ordinal $\langle$cardinal$\rangle$. An *ordinal $\langle$cardinal$\rangle$ operation of type $e \langle C \rangle$* is, by definition, a generalized function F which associates with each set E of ordinals $\langle$cardinals$\rangle$ such that ord $(E, \leqslant) = e \langle$card $E = C\rangle$ a unique ordinal $\langle$cardinal$\rangle$ $F(E)$. An *ordinal $\langle$cardinal$\rangle$ operation of type ∞* is an ordinal $\langle$cardinal$\rangle$ operation which is of type $e \langle C \rangle$ for every ordinal $\langle$cardinal$\rangle$ $e \langle C \rangle$.

For example, let

$$F_1(\{a\}) = a + 1,$$

$$F_2(\{a\}) = \omega_a,$$

and

$$F_3(E) = \cup E;$$

these are ordinal operations of types **1**, **1**, and ∞ respectively. For each set $\{A, B\}$ with $A \neq B$ let

$$F_4(\{A, B\}) = A^B + B^A.$$

Then F_4 is a cardinal operation of type **2**.

Let F be an ordinal operation. An ordinal a is, by definition, *inaccessible for* F iff for every $E \subset a$ such that ord $(E, \leqslant) < a$ [or, if F has type e, every $E \subset a$ such that ord $(E, \leqslant) = e$] one has $F(E) < a$. Inaccessibility for a cardinal operation is defined similarly.

Theorem 69

Let F_i, $i = $ **1, 2, 3, 4**, be the operations just defined.

1^0 The limit ordinals are inaccessible for F_1.

2^0 The critical ordinals are inaccessible for F_2.

3^0 The regular ordinals are inaccessible for F_3.

4^0 The dominant cardinals are inaccessible for F_4.

5^0 The inaccessible ordinals [= cardinals] are inaccessible for F_1, F_2, and F_3; and OGCH for F_4 as well.

We now formulate a new axiom. If X is a set, let $p(X)$ be the set of all subsets of X, as in Definition 1.4.

TARSKI'S AXIOM [AXT]. For every set N there exists a set M with the following properties:

1^0 $N \in M$.

2^0 If $X \in M$ and $Y \subset X$, then $Y \in M$.

3^0 If $X \in M$, then $p(X) \in M$.

4^0 If $X \subset M$, and if X is not equipotent with M, then $X \in M$.

The following two theorems are to be proved *without* using AXC. They may be found helpful in the proof of Theorem 72.

Theorem 70

Let M be a set, and let $S \subset p(M)$ be such that if $X \in S$ and $Y \subset X$, then $Y \in S$. Then there exists a subset N of M such that $N \notin S$ which can be well-ordered *if and only if* there exists a function g with domain S such that $g(X) \in (M - X)$, for all $X \in S$.

Proof outline: Consider the intersection D of all sets K with the properties

1^0 $\{g(X)\} \in K$ whenever $X \in K \cap S$, and

2^0 $\cup L \in K$ whenever $L \subset K$. Let $N = \cup D$.

Theorem 71

Let the notation be that of the previous theorem. If S is equipotent to a subset of M, then there exists $N \subset M$, $N \notin S$, which can be well-ordered.

Theorem 72

AXT implies AXC.

Theorem 73 [*OAXT*]

For every cardinal A there exists an inaccessible cardinal B such that $B > A$.

Proof outline. If $A = \operatorname{card} N$, then $B = \operatorname{card} M$ has the required properties.

The following definition and theorems are generalizations of Definition 4.3 and Theorems 56, 57, 61, and 65.7⁰ which provide additional information on the exorbitant [Hausdorff's term] size of the inaccessible ordinals. Definition 4.10 is in fact the first stage of a transfinite sequence of transfinite processes which push the inaccessible ordinals further and further out.

The technique of Definition 4.10 is a special case of a more general procedure called *derivation*, which can be carried out for any ordinal operation F which is normal. We shall define normality only when F is an ordinal operation of type $\mathbf{1}$; we write $F(\boldsymbol{a})$ instead of $F(\{\boldsymbol{a}\})$. Such an F is *normal* iff it is strictly increasing, and continuous in the following sense: if E is a set of ordinals without a maximal element, then $\cup F[E] = F(\cup E)$. Thus for example, the operation F_2 defined above, is normal by Theorem 60. For each of the generalized functions j_a defined below, its inverse j_a^{-1} is normal; in particular $j_0^{-1} = i^{-1} = F_2$;

j_a^{-1} is sometimes called the *a*-derivative of i^{-1}. We define j_a instead of j_a^{-1}, because it is necessary to show that j_a^{-1} is defined for all ordinals, and, with our particular development, that is best done by showing that j_a assumes all ordinal values, cf. Theorem 74.2.

The terms 'hypercritical' and 'ultrahypercritical' are not standard. The 2- and 3-hypercritical ordinals are sometimes called ζ-ordinals and η-ordinals, respectively.

Definition 4.10

For each ordinal number *a*, the notion of an *a-hypercritical* ordinal and its index j_a is defined as follows by transfinite induction on the well-ordered set $(a+1, \subset)$:

Let *c* be an ordinal such that $c \leqslant a$.

1^0 $c = 0$.

The 0-*hypercritical* ordinals are precisely the initial ordinals, and $j_0 = i$, the aleph index.

2^0 $c = b+1$.

A is *c-hypercritical iff* it is *b*-hypercritical and $j_b(A) = A$. Define j_c as follows:

$j_c(A) = \text{ord}\,(\{B \mid B < A \text{ and } B \text{ is } c\text{-hypercritical.}\}, \subset).$

3^0 *c* is a limit ordinal.

A is *c-hypercritical iff* it is *b*-hypercritical for all $b < c$. j_c is defined as in 2^0.

[Note that 0-hypercritical ordinal = infinite cardinal, and that 1-hypercritical = critical; thus all hypercritical ordinals are cardinals and most of them are critical.]

Definition 4.11

A is *ultrahypercritical iff* A is A-hypercritical.

Theorem 74

Let a and b be ordinals, and let A and B be a-hypercritical ordinals.

1^0 $A = B$ iff $j_a(A) = j_a(B)$, and $A \leqslant B$ iff $j_a(A) \leqslant j_a(B)$.

2^0 There exists a unique A such that $j_a(A) = b$.

Theorem 75

There exist ultrahypercritical ordinals. But if U is the least such ordinal, then neither U nor any smaller ordinal is inaccessible.

[Note that for $a = 0$ Theorem 74 implies Theorems 56 and 57, and that for $a = 1$ and $b = 0$ it implies Theorem 61. Also Theorem 75 implies Theorem 65.7^0.]

Bibliography

This brief Bibliography requires a little comment. Obviously the Bibliography should not be used by students working through this book as a source for proofs of its theorems. In fact the presentation of set theory in this book differs sufficiently from the various presentations in the Bibliography to make such misuse difficult.

The first four books are quite elementary, the fifth and sixth rather less so, while the last three are comparatively advanced mathematical treatises.

1 Fraenkel, A. A. *Set Theory and Logic*.   Reading, Mass.: Addison-Wesley Publishing Co., 1966.

2 Halmos, P. R. *Naive Set Theory*.   Princeton, N.J.: D. van Nostrand Co., 1960.

3 Kamke, E. *Theory of Sets*.   New York: Dover Publications, 1950.

4 Sigler, L. E. *Exercises in Set Theory*.   Princeton, N.J.: D. van Nostrand Co., 1966.

5 Hayden S. and Kennison, J. F. *Zermelo-Fraenkel Set Theory*. Columbus, Ohio: Charles E. Merrill Publishing Co., 1968.

6   Hausdorff, F. *Set Theory*.   New York: Chelsea Publishing
      Co., 1962.

7   Fraenkel, A. A. *Abstract Set Theory*.   Amsterdam: North
      Holland Publishing Co., 1966.

8   Bachmann, H. *Transfinite Zahlen*.   Berlin: Springer, 1955.

9   Bourbaki, N., *Théorie des Ensembles*, Chapitre 3, ASI 1243,
      Paris: Hermann, 1963.

INDEX

m824-1 Tn
57